Karamoko Kourouma
Kaba Kourouma

Termites: A Pacific and Working World

AF154846

Karamoko Kourouma
Kaba Kourouma

Termites: A Pacific and Working World

ScienciaScripts

Imprint

Any brand names and product names mentioned in this book are subject to trademark, brand or patent protection and are trademarks or registered trademarks of their respective holders. The use of brand names, product names, common names, trade names, product descriptions etc. even without a particular marking in this work is in no way to be construed to mean that such names may be regarded as unrestricted in respect of trademark and brand protection legislation and could thus be used by anyone.

Cover image: www.ingimage.com

This book is a translation from the original published under ISBN 978-613-8-47089-2.

Publisher:
Sciencia Scripts
is a trademark of
Dodo Books Indian Ocean Ltd., member of the OmniScriptum S.R.L Publishing group
str. A.Russo 15, of. 61, Chisinau-2068, Republic of Moldova Europe
Printed at: see last page
ISBN: 978-620-4-11975-5

Copyright © Karamoko Kourouma, Kaba Kourouma
Copyright © 2021 Dodo Books Indian Ocean Ltd., member of the OmniScriptum S.R.L Publishing group

PREFACE

I agreed to present this book and this preface so that, first of all, contemporary science would have a place in this series of presentations. As a researcher, I want it to be represented. I would add that I will boldly depart from my area of expertise: I am a mathematician, and I am about to talk to you about the book of a biologist, a naturalist.

This is the second reason, and it has to do with popularization. I think very sincerely and after careful consideration that the advancement of science depends on the improvement of scientific culture. It therefore seems important to me to launch interdisciplinary bridges.

The third reason is rehabilitation. I will try to convince you that this book, which most of you have never heard of, is a *Masterpiece* and that future generations will rediscover it as a fundamental, seminal and programmatic work, because it has contributed to rethinking and reformulating insect biology. Thus, the reader of Karamoko KOUROUMA's book expects to find scientific results, chosen for their impact on biology, for what they modify in our ideas, both about physiology and animal behavior, as well as about the structuring of the biological field.

But the reader, whose culture probably already includes Darwin's *Origin of Species*, as well as Maeterlinck's *The Life of Termites* or Jean-Henri Fabre's *Entomological Excursions*, will also expect to read scenes from the life of insects, which will be like so many incursions into the enchanting and marvellous world of natural history, and which will also fascinate the reader by the more general insights the author will be able to draw on about organisms and their interactions.

I shall confine myself to pointing out the most obvious rhetorical device, that of the modesty of the object studied. From this point of view, insects are a class of animals apparently inferior to more noble species, such as fish or birds, frogs or cuttlefish or cats. In the eyes of neurophysiologists, insects are "a boring subject", which Karamoko KOUROUMA takes care to make exciting. And, among insects, Karamoko is going to look for and somehow rehabilitate the species, not only the most humble, the most hated too, the most misunderstood of insects" but aims at renaturalizing the biological fact

Karamoko KOUROUMA wants to place biological observation outside the laboratory, in nature, thus reviving natural history. He is interested in animal behaviour and its study in ethology. According to him, only this study can help to understand, by analogy, the results of laboratory experiments on the knowledge of animal behaviour.

Naturalist, Biologist, Mathematician and Philosopher, Karamoko Kourouma is a Guinean scientist with a body of work devoted to plants, animals, insects and the exact sciences. In this book, which his family honors me with a preface, he describes termites as social insects and how these beasts feed, organize themselves, reproduce and defend themselves against suppliers, considered their worst enemies.

In general, we have a poor image of termites because they attack the wood in our homes, our libraries and some of our works of art made of wood. Termites are also badly considered because they cause damage to certain crops and forest plantations. In this book, Karamoko KOUROUMA, my teacher, colleague and scientific guide tries to show that the termite is not only this evil animal but that it also brings benefits. This book presents a complete vision of termites. In spite of the damage caused to houses, crops and forest plantations, their ecological role should not be forgotten.

Granted, termites do damage crops, but termites, by working the soil, also appear to help crops. As long as there is organic matter in the soil, it attracts termites, which make multiple perforations in the surface of the soil to come and look for this food from below. All these perforations, up to several hundred per square metre, are so many openings through which water can penetrate the soil instead of running off, thus feeding the water table and limiting the risks of flooding. The soils of the termite mounds are rich in clays that have been brought up from the depths. They are filtering and humid and rich in nitrogen. They are therefore fertile soils that farmers look for to grow the most demanding crops. The work of termites aerates the soil, enriches it and above all facilitates the infiltration of rainwater, which is concentrated in these pockets in which millet or sorghum are sown in certain African regions. This technique makes it possible to multiply yields, especially since the soils are poor, crusted or severely degraded in some Sahelian areas.

I remember that the author was sometimes passive in relation to everything that happened to him (he underwent existence), sometimes dynamic, resisting the demands of life (he created and oriented himself). Would it not be appropriate to see him now only in his dynamic life, this life that I would allow myself to call scientific life, although it seems to me that the desire of this publication would be to show and perhaps demonstrate that KARAMOKO was a man of science and that to rehabilitate him, To pay tribute to him is not only to do a useful work for the deceased and his family, but also to encourage thousands of young Guineans who, perhaps, will one day want to devote themselves to science by following the example of this great patriot and man of knowledge, of know-how and of being. He alone was the living expression of these three dimensions of competence. From then on, a question could be asked: what would be the criterion of appreciation which would make it possible to say if yes or no KARAMOKO was a Man of Science. Thus, by means of our definitions and our observations on the life of KARAMOKO, we will be able to deduce whether or not he was a man of science. Thus, the Universal Encyclopedia, speaking of science, says: "If we admit that natural phenomena obey laws and that these laws are knowable, we can say that science is the whole knowledge of the laws of natural processes". This definition is not very different from the other one which says that "science is an organized body of knowledge relating to certain categories of facts or phenomena. Concerning the criterion of scientificity, the criterion by which it becomes possible or not to say that a piece of knowledge or a work is scientific, the following idea seems to me to be correct: "Any knowledge that has succeeded in inscribing its practices (constructive, deductive, experimental, educational or even foundational), in the context of a scientific approach, would be scientific, Let us then posit that the man of science is the man who, while obeying the obligations of science in his work (objectivity, rigor, precision, etc.), has discovered something useful in his work.The

3

man of science is also the one who, while obeying the obligations of science in his work (objectivity, rigour, precision, etc.), has discovered something useful for his country, or who, failing to make this discovery alone, has participated usefully in the resolution of scientific problems of international interest, either in his country or abroad. The man of science is also the one who, in the face of research, is capable of self-sacrifice, perseverance, rigour and perspicacity. Did KARAMOKO embody all these virtues? Many considerations lead me to affirm that KARAMOKO KOUROUMA known as "KK" was a true researcher.

Apart from my own observations and the benefits of working with this researcher, when on the 40th day of his death, his parents asked Dr Famory KOUROUMA (his cousin), a sociologist at the University of Conakry, to put his affairs in order in Faranah, and on reading this report, I was surprised to see how far the man had devoted his life to research. According to the report of this inventory, the first observations fell on the scientific preparation, in other words, on the physical context conditioning the research work. It is known that this context is generally made up of books, laboratories with everything that could be for the manipulations. At Karamoko, we could list more than 700 books, not counting newspapers and other periodicals. But interestingly, books in Karamoko's library did not have the same life as they have in many living rooms, i.e. to serve as ornaments. The scope of scientific concerns of KARAMOIKO, I would say that he was interested in each of the three fundamental types of science. That is to say: 1-The pure formal type (represented by mathematics); 2-The empirical-formal type (represented by physics) and finally 3-The hermeneutic type (represented by human sciences, psychology, phenomenology, psychoanalysis, sociology). All in great quantity and in an astonishing variety. But of all these works, the most numerous are those of Mathematics, Physics, Chemistry and Biology. The second observation made by this inventory mission concerned its practical equipment or what can be called laboratory equipment, in a context where the environment of scientific research was characterized by indescribable neglect. These are slides in Zoology, Botany, Water and Forests. Three microscopes, -A slide projector, -A small projector of 8 mm films (animated) -One collection of herbaria with more than 4,000 species, constituting the basis of the national herbarium at the Faculty of Pharmacy and Botany of the UGANC. Two jars containing snakes, several cards on Genetics, Agronomy, Geography, Periodic Classification of the Elements (simplified MENDELEV table), Biology, Photo Synthesis, Sex-linked Heredity, Cell Division, the Plant Kingdom etc. The third observation concerned everything that was written by KARAMOKO. In this context and more than elsewhere, it was realized that if he had had a little more time, he would have published a fairly large number of works, including: a-Forty problems for the Baccalaureate b-Inventory of natural resources c-Bush fires d-Rhizogenesis in cuttings e-Tables of normal numbers d-Nimba biology, composed as follows: a first part of the work devoted to 30 problems in biology - An extract from the reading entitled: Philosophy course. An analysis entitled: Entry in the great schools - Collections such as: Science and technique, National colloquium on the plants of popular medicine. Science and literature. - Agriculture: Rice: Volumes 1 and 2 First mission in the mechanized production brigades which was a program of agrozootechnical promotion of the first republic. KARAMOKO had written a lot in other fields, especially Mathematics for example. Annales du Baccalauréat. Mathematics: (course) -Mathesis. Applied Mathematics - 100 practical problems (statements) - 100 practical problems (solutions) - Mathematical logic and various problems - At the end of the novels as "post 5" (published by Harmattan Guinea) - A translation of English: "in the country of Gléglé (King of Dahomey)".

KARAMOKO's contribution to science, education and culture shows that he was never satisfied with reading and writing abstract theories. More than the tranquility of an office, he loved the bush, manipulation, observation in the forests. He walked all over Guinea. At the end of these outings, KARAMOKO congratulated himself by noting that the flora of Kankan; Siguiri, Kouroussa and Kérouané designated as savannah flora is finally studied. This study will be added to those of Beyla, N'Zérékoré, Yomou, Lola, Macenta and Guéckédou. At the time when he was inspector of the Academy in N'Zérékoré, he was much more concerned with nature. With these plants, good results of chemical and botanical analyses were obtained, such as: 1 - One of these plants soaked in an acid solution gave a red coloration and in a basic solution, blue. 2 - The second one gave the opposite results. The importance of this experiment is that it establishes that the barks of these plants could validly replace the indicator papers difficult to convert. By indicator paper one expects in chemistry these papers which indicate the PH of a solution.

Throughout his life as a teacher, his subjects proposed for the various examinations were always retained by the national commission. KARAMOKO, thanks to his constant efforts and to the international conferences and colloquiums for which he was mandated by the Guinean government, managed to make himself known and admired by the great men of science. That is how he found himself representative of MAB in Guinea, member of the UMA (African Mathematical Union) and even member of the World Federation of Scientific Workers.

He participated in the protection of several classified forests in Guinea (Ziama in Macenta) and created a Botanical Garden in Faranah which is today the property of his family thanks to the policy of recognition of intellectual property advocated and implemented by the Government of Professor Alpha CONDE, President of the Republic of Guinea to whom we pay a deserved tribute for his efforts in the promotion of scientific research and the enhancement of the achievements of environmental protection. This original work on termites and termite mounds, fruit of his numerous researches as a naturalist and experienced biologist, is among a series of numerous pending publications which his son Professor Kaba KOUROUMA (Pharmacist-Biologist) and Dr. Jean Amadou KOUROUMA (Mechanical Engineer) have made it their duty to improve, readapt and write for dissemination in newspapers, works and books in order to valorize the priceless works of their late father.

Honourable Professor Aliou Baniré DIALLO, General Coordinator of the Cercle Scientifique Karamoko KOUROUMA (CSKK) Former Minister of Higher Education and Scientific Research

TERMITES: A PEACEFUL AND HARD-WORKING WORLD

According to specialists in animal sociology, termites are the most highly organized of all social insects. But when zoologists give us the identity card of these tiny bugs, a 10th grader who is introduced to the study of invertebrates during the year literally gets caught up in the hierarchy of values. Let's listen to the zoologist: *A termite is a coelomate triploblast metazoan, of the phylum Arthropoda, subphylum Mandibulata or Antennata, class Insecta, subclass Pterrygota, section Neoptera, super-order Blattopteroidae, order Isoptera.*

3 to 5 mm long, with a pale yellow or white abdomen and a brown head with strong mandibles, this is a typical **termite**. It is really a nasty bug whose presence must be reported to the town hall, the library, the museum, as soon as it is detected as it causes damage at high speed.

Termites do not like solitude and live in society like ants, with queens, males and workers. In France According to **the Termite Observatory,** there are now 54 departments infested by **termites** in mainland France. The main regions concerned are the South-West, the Atlantic and Mediterranean coastal departments, the departments bordering the Rhône, Garonne and Loire valleys and finally the Ile-de-F

Termites cause significant damage to buildings: they degrade wood and its derivatives used in construction: plaster, furniture insulation materials, anything containing cellulose is a feast for these insects with an insatiable appetite.

Termites are devious and move in a hidden way, burrowing into the wood from the inside. As a result, the wood weakens and can no longer play its supporting role. In the most extreme cases, this nibbling can lead to the collapse of floors and buildings.

In the invertebrate box, it would be easy to find a termite. **But there are termites and termites**. There are now more than a thousand species of termites in at least twenty genera. And it depends on what they eat. Which can be dry wood, wet wood or rotten wood.

There are cubiterms, cryptoterms, microterms, macroterms, odontoterms etc...

But this line of "terms" tells us nothing about the political, social and economic regime of this underground world.

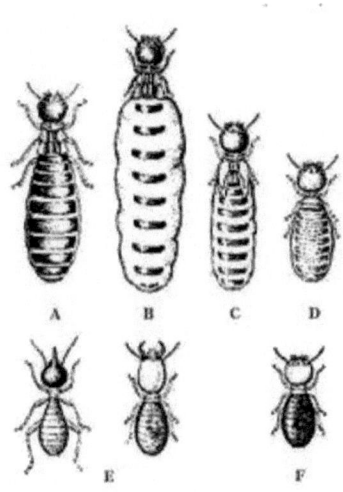

Polymorphism/Polyphenism in termites

A: Main King

B: Main Queen

C: Secondary Queen

D: Queen Tertiary

E: Soldiers

F: Worker

(Polyphenism is the property of animal species that present themselves in several different forms, at the same stage of development or more generally, at the end of their development. It is to be distinguished from **genetic polymorphism** in that the latter depends on genetic variation within a population or species, whereas polyphenism is the expression of different **phenotypes** from the same genetic material (or **genotype**). Examples of polyphenism are a major field of **epigenetic** studies). (source: Polyphenism - Wikipedia)

We know them at least by the damage they cause. In school textbooks, when dealing with insects, students are shown only two types: the cricket and the butterfly. This leads one to believe that the order of insects, which includes nearly four-fifths of the known animal species and of which more than a million types have been described, was made up only of crickets and butterflies. Termites? That's too small, said the teacher. But they are just as big because they live in a perfectly organized society. If we can't dissect a termite, we can at least do a little animal sociology. Without this aspect, what would be the use of science? More than twenty scientists from five continents have studied these social insects: biologists, systematists, entomologists, foresters, sociologists.

Chapter 1: Termitoidea

Termites prefer to settle near a source of moisture, but they can also proliferate on very dry wood.

Interactions of termites with humans

Termites can cause serious damage to human wooden houses.

Ancient manuscript eaten away by termites in a library in Chinguetti (Mauritania)
(Source Wikipedia)

Chapter 2: How termites spread

Sexed winged termites (Wikipedia)

Subterranean termites found in France use mainly two modes of natural propagation:

- **the flight**, generally once a year, of sexual adults (also called swarming), which will leave their colony to found new ones;

- **and layering** (also wrongly called cuttings), which corresponds to a propagation from one person to another via a network of underground galleries of workers sometimes accompanied by other castes (soldiers, nymphs, neotenes). These groups of individuals can be separated voluntarily or accidentally from the original colony and found a new colony.

- A third mode of propagation is similar to cuttings in plants and corresponds to **dissemination** by humans: transport of soil, rubble or contaminated cellulose materials (firewood and construction wood, paper, cardboard, etc.).

Chapter 3. The termite mound: a skyscraper

The art of constructing an ultra-comfortable habitat has belonged to termites for millions of years. And what is more curious is that the material used is as varied as the genera and even the species. This is also true of humans. Earth, vegetable debris and even excrement are used in the most rational way. The cement is supplied by the engineer himself. It is his saliva. With such an amalgam, the termites manage to obtain a real banco outside. This bench can even be as hard as concrete in some termite mounds. For the interior compartments, the structure of the walls is that of pasteboard.

If you look very closely at a termite mound, you will see that it has no doors or windows. This is because termites are destined to live in the dark. The erected part that we see above the ground has only one chamber. In the centre, the royal chamber where the reproductive couple resides. All around is a real garden: the mushroom house and the nursery for the young. In the ground and in all directions, thousands of roads diverge from the termite mound to the supply areas.

Termite mound in Namibia

Creator: Brytta; Credits: Getty Image So it's safe to say that the dome seen above the ground only has the king, queen and young people as tenants. So where are the others? And this is perhaps what makes this society so powerful: the others work and that's all. And work is the supply and defense.

The termite mound is composed of lodges connected by galleries. In species where this caste is present, the workers dig and clean the galleries and collect the eggs laid by the reproductive females (queens or neotenic females). Some soldier termites have enlarged <u>mandibles</u> that allow them to bite their opponents (termites of another species or colony, ants, etc.), while others have a proboscis with which they spray acid at their enemies.

Sometimes, in the fouta, you can see bowls of several hectares literally covered with termite mounds in the shape of mushrooms. These are the dwellings of the cube mounds. One could count at least a hundred of them in a small bowal. As each termite mound houses at least two hundred thousand tenants, a bowal of one hundred termite mounds is a city of twenty million inhabitants. What human agglomeration on earth reaches this astronomical figure?

The average length of a termite does not exceed 1.5 cm. Let's take the maximum belonging to MACROTERMES GOLIATE which reaches 2 cm in length and give this size to all termites.

In the savannah areas you can find termite mounds of 6 meters high and 30 meters in diameter at the base. Built by bellicoterms. Now, 6 meters; that represents 300 times the size of the termite.

A large American building of 300 meters is built by engineers whose average height cannot be more than 2 meters.

Such glory of men is barely more than 170 times their own size. It is a misery!

The longest underground subway in the world is the one in NEW YORK with 29 lines covering 380 km. (This is ridiculous compared to the thousands of underground roads that can total 500 km under a termite mound.

In order to build their skyscraper, the bellisicterms had to move (by assimilating the termite mound to a cone) 2118 cubic meters of earth weighing at least 10,590 tons. The pyramid of Cheops, the first wonder of the world, with its 2,500,000 cubic meters weighs 6 million tons.

If one grasps a rigorous comparison. One realizes that the termite skyscraper is the true wonder of the world. These are sobering figures about the misery and splendor of mankind in relation to the animal world.

Every modern home needs water, light and air conditioning.

For water, termites go to the water tables they know.

For light, they don't need it. Since they exist, they are under the ground and are well there.

Air conditioning is another problem.

The aerial part of the building is occupied as we said by the royal chamber, the vegetable garden and the young people. This garden is cultivated with mushrooms. And as far as we know it is mainly mushrooms of the family of rylariaceae. These mushroom beds communicate with the outside world through small openings. All in all, these gardens are nothing more than air conditioners. They serve to condition the air in the nursery and the royal chamber. However it is necessary to see in the cultivated mushrooms a direct or indirect source

Termite mounds are worth their weight in gold *(by Dominique Raizon Article published on 05/02/2008 Last updated on 07/02/2008 at 17:26 GMT)*

The Djenné mosque in Mali. (Photo: Claude Verlon/ RFI)

A new science, bio-mimicry, is inspired by the natural functioning of ecosystems to solve human problems. Thus, for example, architects take a close interest in the way termites construct their buildings and gold miners rely on the geological indications provided by termite mounds (Maximilien Quivrin, entomologist)

Chapter 4. INTEGRAL SOCIALISM

Sociologists refer to the division of labour within termites as castes. It is perhaps that the social organization itself. And this is the proof of the solidity of this regime. In spite of its feudal consonance, the word caste will therefore be retained with regret, because in the society of men it would be aberrant to speak of the caste of teachers or sailors. What is clear is that termites are unaware of the struggle of glasses. They have reached the state of integral socialism long before men, even when they were born on the earth. To say, as some philosophers affirm, that a society without love is a termite mound is, as we shall see, the biggest human lie.

The termite society is hierarchical in the simplest way: either you have a sex, or you don't, or you have a little. Then, either one is a monarch, or one is a soldier, or one is a worker, or one is a king

The sexed ones are the king and the queen.

The neutrals, i.e. the non-gendered, are the soldiers and the workers.

The nektonics are those who have little sex. The nature of this vague sex will be specified later.

Among the soldiers, there are small and large. They are called minor and major. The same is true of workers. When a termite is born, nothing distinguishes it from its brothers and we cannot say if it will be a soldier or a worker, a king or a queen? All the differentiations will take place in the course of moults.

To moult is simply to change clothes. Snakes also have fun moulting. At a given stage, the little larval termite gets rid of its envelope that has become too tight and makes another one. Now, during these successive moults, the castes will separate. But it is not because one has moulted that one becomes a soldier. It is because during the growth, between two moults, one has been nourished in a certain way or more precisely because one has been in a certain way. First conclusion, it is food that makes the soldier or the worker, the king or the queen.

As soon as the green light is given to the muo, the first such operation separates those with a large head from those with a small skull. Now the big heads are future soldiers or workers, the small skulls are future kings and queens as well as nektonics.

At the second moult there are not many changes. Each one keeps its physiognomy while improving it all the same a little

At the third moult, the soldier and the worker separate. The soldiers have a very elongated head and the workers have a round head. Second conclusion: the habit allows to recognize at least the monk

At the fourth moult, the worker and the soldier stop there. They have no sex. They are neutrals

The other two castes continue the parade.

At the fifth moult, the king and queen emerge. They are the only sexed ones, around them, termites that are neither king nor queen, neither workers nor soldiers. They have no sex, but yet they must have it. It is because their sex is only in incubation, in a latent state, incomplete, ready to emerge at the first opportunity. They are the neotenics. We will see their role.

So we are in front of all the castes after the 5th moult. The habitat being built and the society in perfect balance, what roles will the different castes play?

The king and queen have a very precise schedule: they have to produce twins without interruption. The queen will therefore lay eggs all her life.

The workers, both large and small, are responsible for the food supply of the whole society. They are in charge of the breeding of the young, the maintenance of the garden and the repairs to be made to the termite mound. They are completely blind, and some of the workers are even known to be termites, and they do not even have a trace of optic nerve.

The soldiers, who are characterized by an elongated head ending in mandibles armed with sharp teeth, have the role of defending the community against aggressors. They are unable to feed themselves.

The neo-tonics, these dubious sexed losers seem to come from a transitional nude between the 4th and 5th are intended to replace the king and queen in case of death of the royal couple

When it comes to the division of labour, you can't find a better organized society.

Chapter 5. THE QUEEN: A LAYING FACTORY

The priority problem in such a community is certainly that of supply.

Feeding the king and queen, feeding the soldiers, feeding the neo-tonics, feeding the young, all of this is the work of the workers. Soldiers already represent 15% of the total population. This is a heavy burden.

But feeding the queen seems to be the most serious occupation of the workers. While an adult termite is 1 cm long, the queen may be 10 cm long. But that wasn't all. Its volume has taken on proportions so disturbing that its abdomen is two thousand times bigger than its head. This monster therefore constitutes a real attic. But as its function is to lay eggs, it must eat the most to lay the most. One day, we attended the projection of a magnificent son on termites at the Polytechnic Institute's banquet hall. We saw a cohort of workers stuffing the meal into their mouths. At the other end of the same queen, another army of termites were rushing to the eggs that were coming out at the rate of 20 per minute. A queen can lay up to 10 MILLION eggs a year. Fortunately, every year emigration brings the population back to more reasonable levels. The amount of raw material needed to run such a biological factory is not likely to be small.

The workers carry the eggs in blankets. When the young are hatched, it is necessary to die, soldiers and royal couple are almost fed with bottles. The workers regurgitate in the mouths of the soldiers a paste made of chewed cellulose fragments mixed with saliva.

As for the young, the workers feed them only with their saliva.

The activity of the workers in a termite mound is visibly restless.

Everything in this community seems to focus on two important matters: laying eggs and eating. But eating what? Eating wood, i.e. lignin, i.e. cellulose. This is perhaps the first question that nature has asked of living beings. How to eat cellulose.

Chapter 6. THE INSTESTINE OF A TERMITE: A FAIR OF BACTERIA

Termites eat wood. And this is what gives them their bad reputation. For the two hundred million years that they have existed on earth, these beings have eaten wood. Ruminants that live on grass and other plants also live on cells. In the classroom, children are taught to draw a diagram of the digestive tract of a ruminant, but they are hardly told how this large tube called the rumen works.

Digesting a starch molecule into glucose is not a difficult problem for humans when their digestive tract is working at full capacity

But breaking down a molecule and making thousands of glucose molecules is a job that no vertebrate can do. This task must be entrusted to those aberrant organisms that are bacteria. Since in the forest, these bacteria are capable of decomposing plant matter, there was nothing left to do but to mobilize them for their own purposes. This is what ruminants have done. Their digestive tract is lined with billions of bacteria capable of breaking down cell molecules and making them available to beef.

This is not free work, because the bacteria find in their host all the conditions of comfort to multiply. But since these single-celled creatures have the particularity of reproducing exponentially, if the ox were to let them play freely, the poor ruminant's belly would shatter. So, a ruminant always finds a way to put a stop to this disturbing growth of bacteria. It digests them.

This ox-bacteria life is a symbiosis, about which there is much to be said. Ruminants have thus answered the problem posed by nature. Unfortunately, it's a stolen patent. They stole it from the termites.

Ruminants are not new in the scale of animal paleontology. Termites are at least 200 million years old. But they knew the technique of colonizing bacteria on their own account. The digestive tract of termites is a veritable bacteria fair. Cellulose is processed with a yield of around 90%, while cow dung is full of plant debris. All this is a simple symbiosis. Termites are capable of inventing other things.

Professor GASSE, who knows these beasts well, explained the real reasons for the gardening that termites do.

Termites of the genus Macro terms make pellets of chewed wood. These pellets constitute a favourable environment for the development of fungi. So we have mushroom moulds. Mycota thrive in these pellets. The fungi consumed by the workers probably provide them with certain growth vitamins.

But fungi serve another, more serious purpose. They attack the lignin of the wood and release the cellulose. And according to GRASSE, they even seem to modify the cellulose, the termites eat the very material of the modified millstone and continuously bring in new material. The ingested materials are attacked by cellulolytic bacteria living in the termites' hindgut. This is a double symbiosis.

First symbiosis: fungus and termite live in mutual benefit.

Second symbiosis: the bacteria and the termite live off the material adopted by the ruminants. For the termites, the supply of vitamins and carbohydrates seems to be regulated by this double symbiosis.

The protein has yet to be found.

The rectal cavity of termites is another fair: a fair for flagellated protozoa, many of which phagocytose wood particles. They cause the wood to undergo a transformation of cellulose into glucose, which, through fermentation, gives carbon dioxide, hydrogen and acetic acid. But this is not yet the important part of the problem. The termites, by exchanging saliva and rectal fluid with each other, supply each other with flagellated protozoa, which are the main source of protein. So the protein is found.

Chapter 7. IN THE GUT OF A TERMITE, A BIOLOGICAL MONSTER: MIXOTHRICA PARADOXA

The gastronomic bacteria of termites, from single to double symbiosis, thus appear to be varied. But it gets better. Termites host a protozoan that is almost a biological monster. In the gut of some Australian termites, a curious flagellated protozoan with the barbaric name of mixothrcaparadoxa was found. As they had too many flagella, it was already put in the multi-flagellate drawer. However, it was later discovered that the animal did have flagella, but that not everything that was called flagellate was flagellate.

The sociologists removed it from the case paper and put it in the flagellates' case, as a temporary measure. They studied this transient protozoan further. They found that the flagella covering the body are actually spirochetes, shaped like elongated bacteria. There are large and small ones. Worse than that, each spirochete is associated with an ordinary bacterium. And to confuse matters, other symbiotic bacteria live inside the animal. It's a complex symbiosis that ruminants are not capable of.

Chapter 8. NATIONAL DEFENCE

So far, we know the role of the royal couple and the workers. But what do the soldiers do?

First, we know that they are unable to feed themselves. Their intestines do not harbour bacteria or protozoa. At their age, they are still bottle-fed.

Their only usefulness lies in their mandibles which are terrible weapons of war. Their task is to defend the termite mound against possible aggressors.

As hereditary enemies, termites know only one: ants.

We must admit that ants are probably the most quarrelsome insects on earth. Termite and ant are however of the same class. But one belongs to the order of isoptera and the other to that of hymenoptera. So it is not a struggle of classes, but of orders. Now, as much as ants like fighting, termites like peace. We have never seen termites attack another termite mound. They are always ready to defend themselves and that's all.

The national defence weaponry held by the soldiers is quite sophisticated.

First, there is the white weapon. The soldiers who use them have mandibles with sharp blades that function like shears. An ant seized by a termite is literally cut in two.

Then there is the chemical, rather bacteriological weapon. The soldiers who use this bomb have a nose ending in a tube that communicates with a cavity in the skull and is filled with a viscous liquid. It is this liquid that the fighter projects violently on his enemy. An ant bombarded by this jet is lost. The glue immobilizes it. If it struggles, however, it's only to become even more engulfed. It dies of asphyxiation.

What do these soldiers do at their battle stations. They stick out their mandibles or their long noses through the openings while keeping their abdomen inside the termite mound. This is because the abdomen, which is not armoured like the skull, is the most vulnerable part of the body.

The tactic is to never fight in open terrain because, if the termites' arsenal is quite modern, the ants' is even more sophisticated. Having completely blocked an opening with its abdomen the termite waits.

Let an ant pass by. It is immediately cut in two pieces by a soldier with a knife or bombarded with chemicals and engulfed.

If an ant manages to break the siege around the fort by entering the termite mound, it has committed suicide. It is immediately surrounded by a swarm of soldiers who soon tear it to pieces.

It is really rare that the victory does not go to the termites. In general, the ants, after many reinforcements with the most sophisticated weapons, after a siege in good standing, never succeed in defeating the defenders. So, they retreat, in front of these retarded of the sociological scale coming behind them by at least twenty orders in the coherent classifications. They do not bury the dead.

But war cannot be continuous. Soldiers have to do something in peacetime. They keep watch. They are vigilant and their very presence is something that drives away anxiety. However, there is an inner order to be maintained in the termite mound. It is just likely that the maintenance of order is devolved to the soldier miners. When the workers are carrying the eggs from the egg-laying factory line, the little ones that circulate around the workers seem to be enforcing an order, for in any society there are pests.

But a burning question arises: at the rate of operation of the factory, a frightening number of soldiers is to be expected, constituting a heavy burden for the workers. How to plan the birth rate of those who cannot feed themselves?

Since the type of food given to the larvae directs the termite towards this or that caste, it would be enough to act on this parameter to stop making soldiers. But it is better to have a young soldier than a tired one. So we act on another parameter. These brave fighters in excess are easy to eliminate. You just have to stop feeding them. And that is what happens in a termite mound. The reduction of the defense budget automatically regulates the birth rate.

Chapter 9. PILGRIMAGE

The elimination of tired soldiers is not enough to maintain a reasonable rate in the city since births continue. It is necessary that some emigrate. This happens once a year. But who are those who emigrate?

Among the youngsters in the nursery, there are a certain number of them oriented towards reproduction each year. They will be true breeders, quite different from the nektonians. These princes with dubious sexes. These new breeders are destined to be actual kings and queens.

When the time of departure approaches, these newly chosen ones adorn themselves with two pairs of wings. Then one day, especially at the beginning of the wintering period and after a rain, one sees swarms of winged termites in the sky. If the flight takes place during the day, the birds cut themselves a share, even a large share, in the cloud. Some lucky ones make it to the ground.

On forced landing (because these wings cannot hold the air for long) the termite easily gets rid of its cumbersome car and starts wandering.

In this walk, if it meets another opposite sex, the pilgrimage ends, this is what we easily observe two termites following each other as if chained by a sexual fluid, the female in front, the male behind, the one in front stops dead. More than a sexual fluid, one has the impression that a material thread links two congeners.

They will enter a hole, preferably located at the foot of an old stump.

All the winged termites that we see in the air at a certain time of the year are sexed termites.

These two termites, which by chance have met, will be the origin of a new termite mound. However, if the queen starts to lay eggs, ready-made workers will be needed, but she has made the king and queen capable of fulfilling this role.

The queen will continue to keep her pilgrim appearance but not the appearance of a monstrous queen commanding a growing kingdom. The royal couple takes care of all the breeding work as well as the first foundations of the future city.

They are the only two workers present.

When young people reach working age, they take over from their parents, who are now only concerned with egg production. Little by little, the termite mound takes shape and the community is established.

Thus, new termite mounds are established every year.

Fortunately, the birds are there to make a reduction. But also, a termite that has lost its wings and is wandering around in the wild may meet an ant. In that case, its account is settled. And there are other terrestrial enemies. Other termites, carried by the wind, will cling to trees and never meet a partner. All of these factors contribute to reducing the number of lucky pairs to an extreme.

Chapter 10. THE TWILIGHT OF A TERMITE MOUND

It happens in the royal couple, that a spouse dies. The first one as a result of all the efforts made in childbirth. One must be astonished rather to see her alive having released a million eggs.

In this case, the nektonics, these princes with temporary sex take over. Among them, the king will choose, not usually one, but three or four wives. This transition from monogamy to polygamy is the dawn of a real anarchy.

With four queens manufacturing eggs for the queen, the termite mound will soon become abnormally overcrowded. What's worse is that male nektonians will start competing with the king by choosing wives for themselves. In this mess, there is no longer any thought of feeding the army. Without soldiers, the termite mound succumbs to the first ant attack and the colony dies out for lack of defenders.

Monogamy had kept the colony together for years, but an unusual polygamy gave the green light to unbridled debauchery among the princes, who sowed intolerable disorder in the termite mound.

Professor GRASSE has shown in the course of experiments that 50 termites are necessary to create in a colony of this size a certain group effect which coordinates activities. It seems that next to this lower limit, it is necessary to place an upper limit beyond which there is anarchy. This upper limit can be exceeded precisely by the conduct of these wanton nektonics who seem to refuse a cynical measure: to keep indefinitely within themselves a fictitious sex that serves no purpose.

The most peaceful society of the insect world, after periods of prosperity and military glory, ended up in a decline and a definitive death initiated by a disturbance in the matrimonial regime and completed by a real coup d'état perpetrated by nectonic people with vague sex.

Chapter 11. TERMITES IN THE DOCK

Sneaky as a termite, that's the biggest insult you can throw at a human.

A million-volume library crumbles to dust. We're looking for a culprit. It's the termites.

A ceiling collapses on the skulls of the tenants in a wooden house, we look for a culprit. It's the termites.

A mud wall deflates like a punctured balloon and buries quiet sleepers. We look for a culprit. It is the termites.

These are common scenes that at least warrant a review of the criminal record of the most peaceful insects on earth.

The deviousness of termites is matched only by earthquakes. They always present themselves not as ready to devastate, but as having completed the devastation. To say that they attack only dead wood is to forget the immense damage they can do to cuttings and even to standing plants.

The destruction of cuttings by these insects seriously compromises an entire agricultural campaign.

Young dead plants in a nursery are the first to be attacked by termites if the environmental conditions are favourable. And in this case, these insects do not hesitate to switch from dead to living plants.

Chapter 12. TERMITE CONTROL MEASURES

How to spot the presence of termites?

Termites are not easy to spot. The most frequent signs of infestation are tunnels raised from the ground called "**termite** cords". They appear on the surface of walls, wood or soil and are used by the insects to supply their colony with cellulose.

Another sign of **termites** is to find sawdust-like powder near cracked or brittle wood.

A final symptom is that **termite** droppings can take the form of small pellets that are usually found near the wooden fixtures of a house.

If in doubt, knock against furniture, beams and frames to see if they sound hollow or test their porosity with a spike. If it sinks easily into the wood, it is best to contact a professional immediately.

How to exterminate termites?

These insects can only be killed with strong chemicals.

Termite detection traps are first placed every 2 or 3 m on the ground to check for their presence. They have the double advantage of eradicating the colonies while preserving the building. If the presence of insects is confirmed, baits containing an insecticide with a delayed effect are used. **Termites** ingesting the insecticide spread it before dying in the termite mound. Inside a house, the **treatment** often involves an **injection** carried out in three stages: drilling, installation of injectors, and then **injections** allowing the curative biocidal product to be deposited inside a material.

Masonry cellar floors require a line of **injection** by drilling vertical wells (in line with the walls). On dirt floors, insecticides must be applied to the entire surface of the floor.

For the **treatment of** perimeter walls and partitions, the **injection** line is carried out by drilling horizontal wells in the walls (as close as possible to the finished interior floor). The buried walls are treated in a grid pattern, with horizontal wells drilled over the entire height of the buried walls.

For structural wood, the **treatment** goes up to the level above the infestation with a double **injection** at the recesses. All structural wood in contact with the masonry is treated by **injection along** its entire length. This **treatment** ends with a double surface application.

The price of an antitermite **treatment** varies according to the size of the insect colonies: the **treatment** ranges from 1500 to 3000 euros depending on the size of the roof and the degree of infestation. In case of heavy infestation, the cost of the **treatment** can reach + or - 50 euros/m2.

Termites like humidity: to avoid their return after **treatment, it** is advisable to periodically check the humidity of the areas at risk and ventilate them as much as possible.

G. SCHNTZ, of the division of phytopathology and agricultural entomology in Zaire (now DRC) has studied against years, what it is appropriate to implement to fight against termites.

The first way is direct control, which consists in destroying the termite mounds. But by levelling the ground, it happens that this result does not pay off because most termites are subterranean and we know that among them, nectonic beetles are ready to replace the royal couple. We must therefore keep an eye on the destroyed termite mounds and raze them again as soon as a semblance of a mound appears.

A second means of direct control is DDT vapour, which has given excellent results. Preventive methods are certainly more effective against these insects, which generally only appear after damage has occurred. Several methods have been tried with varying degrees of success

In crop protection :

> Preventive soil treatment using chemical compounds

> Preventive treatment of seeds and cuttings coated with DDT powder

> Treatment of standing plants by applying H.C.H. powder to the collar

In the protection of buildings :

- Ground treatment
- Termite-proof *construction*

In wood protection

❖ Chemical treatment of wood

❖ Use of suitable wood.

These last two points having a direct economic importance, we give below two tables of C. SCHMITZ, obtained after several tests.

DURATION OF WOOD PROTECTION ACCORDING TO THE TREATMENT

PRODUCTS		ACTIVE MATTER(%)	DURATION OF TOTAL PROTECTION
	I. AQUEOUS SOLUTION		
1	Copper sulfate	1	4 years
2	Zinc chloride	1	4 years
3	Ferric chloride	1	4 years
4	Bromide silicifloride, calcium acetate	0,5	3-4 years
5	Diphenil- mercury	1	3-4 years
6	Phenil-mercury chloride	1	3-4 years
7	Pyridyl mercury stearate chloride	0,2	1 year
8	Lead or mercury Naprenate	4	2 years
9	Calcuim, zinc and copper Naprenate	12	/
10	Sodium zinc lead pentachlorophenate	0,2	3 to 4 years
11	Thenite(isobornyl thio cyanate)	2	400 days
12	Zinc or copper dimothyl-dithiocarbamite	0,5	500 days
	II. Solution of organic solvents		
13	Copper pentachlorophenate (alcohol)	0,4	3-4 years
14	Xantone (acetone)	0,5	3 years
15	Ferrous dimothyldithiocarbamate.	0,5	3-5 years
16	Pentachloropheno l	1	11 months
17	Pentachloropheno l	2	4-5 years old
18	D.D.T	1	1 month
19	D.D.T	2	3-5 years
20	H.CH gamma	0,01	4 months
21	H.CH gamma	1	At least 1 year
22	chlordane	0,5	2 months
23	chlordane	1	At least 1 year

RESISTANCE OF WOODS TO TERMITES

Common names	Scientific name	density	Duration resistance	Toxic principles for termites
Ronier	Borassures ethiopium		Refractory	
	Afzelia africana		4 years	Toxic
	Afzelia pachyloba		4	Toxic
Iroko	Chlorophoraexcelsa	0,6-0,7	7	chlorophorin
	Cynometra		4	tectoquinone
Tali	Erythrophleumguin Fagara	09 0,8-0,9,	6	
	macrophylla Piptadenia		5	
	griffoniana		4	
Daboma	Piptadenia africana	0,6-0,8	4	
	Pterocarpus soyauxii		4	
	Pterocarpus angolense		4	
	Pterocarpus tinotorius Staudtia		4	
	camerunensis		4	
	Staudtias tipitata		4	
Teak	Tectona grandie		4	Tectoquinone
Azobe	Lophi reprocera	1,1	6	
Ovala		0,9-1,1	6	
	Pentacle thramacrophyl		Immunized	
	Hymenocardi ascida		Immunized	
	Terminelia glaucescons		Immunized 2-5	
	Dichrostachys Combretum		years 2-5 years	
	Bridelia micrantha dalbergia		2-5 years	
	melennoxylon			
	Lasiodiscus midbrandii			
	Sensitive plants			
Okoumé	Aukoumeaklaneana			
	Entandrophregma	0,5-0,6		
mahogany	Khaya	0,5-0,6		
	T erminaliasuperba			

BIOGRAPHY, RESEARCH, WORKS AND LIFE OF THE LATE KARAMOKO KOUROUMA

*The late Karamoko KOUROUMA, a self-taught **naturalist, biologist, botanist and mathematician, was born in 1924 in Baranama, Kankan Prefecture, into a farming family that also held a political office, that of the customary chieftaincy.*

*His father, **Fakoly KOUROUMA**, served in Kankan, from 1925 to 1942 as a representative of his elder brother Dininba KOULOUN, chief of the canton of Sabadougou. It should be noted that **Fakoly is the honorary name often given to the KOUROUMA**, especially to those who would have made some brave actions. Also, the real name of KARAMOKO's father is KABA (my homonym).*

*Moreover, Karamoko is called Fodé, a religious designation of the kind of Thierno in Fouta. KARAMOKO received this name in testimony of the respect of his parents for a great marabout who lived with them when KARAMOKO was born. If we recapitulate, we will find that KARAMOKO is called **KARAMOKO FODE KOUROUMA***

His paternal grandmother Dinimba KONATE is from a small village, Batiguila, in the sub-prefecture of Baranama. This grandmother had only three children:

1- *Dininba Kouloun or KOUROUMA Ibrahima who, after Kaba Kouloun, will be chief of the canton of Sabadougou from 1912 to 1946.*

2- *Facoly of his real name KABA (Father of KARAMOKO)*

3- *Sonafing*

*On the strictly paternal side, KARAMOKO is from a large family. His father had married 4 women: first Gneriba KEITA who had 4 children; then Sogbè CONDE who had 4 children; **thirdly Adama CONDE mother of KARAMOKO. She had 6 children of which Karamoko was the first**. They are successively KARAMOKO, Poret, Wamba, Magnini, Dinimba and Nfaly (no survivor in 2018);*

KARAMOKO had long rejected any idea of marriage. In spite of everything, he made a child with a certain Oumou DIALLO, this child Ibrahima KOUROUMA was a driver in Ivory Coast and died in Conakry in 1990 following an illness. It is thereafter and on recommendation of the chief of the village, without mutual consent that he married **Fatoumata KEITA** *(Daughter of DOUA KEITA a Chief Warrant Officer in the pay of the colonizer and Kensa KONATE) who made him 8 children namely:*

1- **Jean Amadou KOUROUMA,** *is the homonym of the teacher Mr Jean CAMARA of Forécariah, who voluntarily supervised and supported him. Jean Amadou finished the university of Conakry and a doctorate in USSR in Machine-Tool (Technical Engineering).*

2- **Kaba KOUROUMA,** *homonym of his father, did a lab-pharmacy at the University of Conakry, a Master in Public Health (MCM) in Antwerp/Belgium, a DES in Hemobiology and a University Diploma in Immunohematology-Transfusion in Abidjan. He is currently working at the CNTS in Conakry.*

3- **Adama KOUROUMA** *who studied at the University of Kankan, Physics option, is currently working at the Ministry of National Education*

4- **Daouda KOUROUMA** *who is a Civil Engineer and lives in Russia.*

5- **Aicha KOUROUMA,** *secondary school computer teacher for the Ministry of Education*

6- **Sarangbè KOUROUMA,** *Pharmacist, manages the Pharmacy K KOUROUMA in Kenieen/Conakry and is a member of the National Council of the Order of Pharmacists of Guinea.*

7- **Mamadi KOUROUMA,** *is an electrical engineer in Malabo (Equatorial Guinea)*

8- **Bintou KOUROUMA** *, the last one is a sociologist and lives in France :*

In reality, Fatoumata KEITA had 9 pregnancies with 8 living children. In row, the third died following an accident of transfusion.

At the school age, KARAMOKO enters the primary school of Kankan by the care of DOUA KEITA, who in these years, was a very influential circle guard. It is by friendship for the father of KARAMOKO that DOUA, for already long years, had asked KARAMOKO in education.

But, Mr DOUA KEITA, father of Mrs KOUROUMA (Fatoumata KEITA) also had a teacher friend named JEAN CAMARA (Métis). He lived in FORECARIAH. Because of the long and good friendship of the two men, Mr Jean had asked his friend to give him one of his sons to educate him; this is how KARAMOKO will be designated.

KARAMOKO will thus make his primary studies a little everywhere with Mr Jean until the certificate of Primary study passed in NORASSOBA. Then he was admitted to the Georges POIRET school, then to the FREDERIC ASSOMPTION Normal School in 1942 with the ^{1st} classification out of 50 (Water and Forest).

In 1954, at the end of his studies, he was admitted to the ^{1st} part of the graduation diploma of the Ecole Normale: ^{1st} out of 46 students, he was admitted to the ^{2nd} part of the diploma of the same school (Water and Forest session). The same year, he was awarded the prize for the best student of the General Government.

He is appointed to the common secondary cadre of forestry assistants of the A.O.Fen as a 6th class assistant, trainee.

In 1947, he entered the ^{second} year of the Forestry School as an assistant.

In 1948, he was admitted to the exit exam of the Forestry School and posted to DAHOMEY (now Benin). The Government of DAHOMEY decided to assign him to COTONOU.

In 1950, he was integrated into the transitional hierarchy of the senior common cadre of forestry assistants.

In 1952, he was assigned to NATITINGOU, he was appointed on August 14, 1952.

In 1953, he was posted to PARAKOU.

It should be noted here that from 1948 until the end of his life, Karamoko walked two paths: on the one hand, he worked as an administrator, and on the other hand, he continued his studies.

Thus, on June 17, 1953, he was granted a 12-month leave of absence at his request. In July of the same year, he left PARAKOU for Kankan.

Having obtained on August 9 of the same year an authorization to go to France, to prepare in PARIS the first part of the baccalaureate, Modern series. He will obtain this first part of the baccalaureate on July 13, 1954. KARAMOKO prepared this first part under the control of the Free University Institute of PARIS (Boulevard Saint Michel).

In November 1954, he enrolled at the Universal School by correspondence for the preparation of the second part of the Baccalaureate, Elementary Mathematics series. In June 1955, the Director of this school said of him: **"the student has worked and made progress especially in science; overall, the success would be justified".**

Thus, on November 3, 1955, he was admitted to the 2nd Part: Elementary Mathematics.

From this second part of the baccalaureate, the life of Karamoko will take another direction than that of the Water and Forests.

According to his wife, it all started with a small incident.

Indeed, as soon as he was admitted to the baccalaureate, KARAMOKO would have asked the Water and Forestry authorities to integrate him into the senior staff of the Water and Forestry Service. The answer to his request was negative. It was then that he decided to leave this body. He worked hard and succeeded in joining the teaching corps.

Thus, in terms of work at the Water and Forestry Department, he successively held the following positions

ABIDJAN in 1946, BOUAKE, in 1947, COTONOU in 1949, NATITINGOU in 1952, PARAKOU May 1953. On July 28, 1954, by decision n°5508, he was assigned to

Guinea (see official journal of 7/8/54. The decision n°704/DF of 10/2/55 appoints him as the manager of the revenue office in Macenta.

Having obtained his resignation from the Water and Forestry Corps, KARAMOKO went to teaching by decision n°9212/E/9 of September 30, 1957 and took up his post on October [1,] 1957. He worked in N'Zérékoré as a primary school teacher in October 1957. Then he was assigned to the complementary course of LABE where he stayed until July 14, 1958.

*The same year, he was assigned to the primary school of Baranama, his native village, for one month and 10 days. Then he returned to the complementary course of Macenta which received him on 17/11/1958. He remained in Macenta until 1962, when he was victim of a traffic accident, which immobilized him for eight months in hospital. It is known that KARAMOKO limped a little. He was transferred in 1963 to the academy of Kankan as Director of the [2nd] and [3rd] cycles. He stayed there until 1967, when he was appointed Director of [2nd] and [3rd] cycle at the Academy of N'Zérékoré. **On 30 April 1969 by decree n°130/PRG he was appointed Director of Research at Mount Nimba**. He did not stay long in this position because in 1971, he was appointed, in spite of himself, Regional Director of Education of N'Zérékore and in 1972 Inspector of Academy of Forest Guinea. This long stay in the forest will have a great importance in his life. He would have liked to remain in this area when, in **1973, he was appointed General Administrator of the IPGANC, now the University of Conakry.***

The Administration: what a horror for KARAMOKO. That is why, while his wife did not expect it at all, KARAMOKO asked for his transfer to Faranah (synthesis of three climatic zones: Forest, Foutah and Upper Guinea). He will be posted there as Inspector of Academy during the school year 1975/1976 and will remain there until 1980.

One day in 1980, Karamoko went to give a conference on the theme "tobacco or health, you choose" at the Faculty of Agronomy of Macenta, from where he returned sick. He went to Conakry and then to Romania for treatment. On Monday 28 July

1980 he died in a big hospital in Bucharest, the capital of Romania.

It should be noted that KARAMOKO, because of the many stresses of his job and the many nights spent reading and researching, smoked and drank a lot of coffee.

One thing is the will of men, another is the will of God.

We have just seen KARAMOKO in his birth family, in his own family, at school and in the administration, i.e. his staticodynamic life.

In this life, in fact, man is sometimes passive in relation to everything that happens to him (he undergoes existence), sometimes dynamic, resisting the demands of life (he creates and orientates himself). Would it not be appropriate to see him now only in his dynamic life, this life that I would allow myself to call scientific life, although it seems to me that the desire of this symposium would be to show and perhaps demonstrate that KARAMOKO was a man of science and that to rehabilitate him, To pay homage to him is not only to do a useful work for the deceased and his family, but moreover, it is to encourage, at the same time, a thousand young Guineans who, perhaps, will want to dedicate themselves to science one day by following the example of our illustrious departed.

From then on, a question could be asked: what would be the criterion of appreciation which would make it possible to say if yes or no KARAMOKO was a Man of Science.

To find the answer, let us adopt the position of briefly defining what is science and what is a man of science without taking into account KARAMOKO. Thus, by means of our definitions and our observations on the life of KARAMOKO, we will be able to deduce whether or not he was a man of science. Although there may be particular definitions of science depending on the type, they all agree on certain ideas common to science in general.

Thus the Universal Encyclopedia, speaking of science, says: "If we admit that natural phenomena obey laws and that these laws are knowable, we can say that science is the whole of the knowledge of the laws of natural processes". This

definition is not very different from the other one which says that "science is an organized body of knowledge relating to certain categories of facts or phenomena.

With regard to the criterion of scientificity, the criterion by which it becomes possible or not to say that a piece of knowledge or a work is scientific, the following idea seems to me to be correct: "would be scientific any knowledge that would have succeeded in inscribing its practices (constructive, deductive, experimental, educational or even foundational), within the framework of a regulated set of operations, that is to say, of transformations governed by formal schemes"

Let us then posit that the man of science is the man who, while obeying the obligations of science in his work (objectivity, rigor, precision etc.) discovered :

a- *Something of a tool for his country*

b- *Who, failing to make the discovery alone, has made a useful contribution to the solution of scientific problems of international interest, either at home or abroad.*

The man of science is also the one who, in the face of research, is capable of abnegation, perseverance, rigour and insight.

Did Karamoko embody all these virtues? Many considerations lead me to affirm it.

KARAMOKO KOUROUMA was a researcher

When, on the 40th day of his death, his parents asked me to put KARAMOKO's affairs in order, I was surprised to see how far the man had devoted his life to research.

My first observations fell on the scientific preparation, in other words, on the physical context conditioning the research work. We know that this context is generally made up of books, laboratories with everything that could be for the manipulations.

At Karamoko, I was able to list more than 700 books, not counting newspapers and other periodicals.

But interestingly, the books in Karamoko's library did not have the same life as they have in many living rooms, i.e. to serve as ornaments. On the contrary, I noticed that they were all regularly used, each one according to its day.

Before quoting some titles of works and also in order to allow you to have an exact representation of the extent of scientific concerns of KARAMIKO, I would say to you that he was interested in each of the three fundamental types of sciences. That is:

1- The pure formal type (represented by mathematics) ;

2- The empirical-formal type (represented by physics) and finally

3- The hermeneutic type (represented by the human sciences, psychology, phenomenology, psychoanalysis, sociology)

You could find a little bit of everything at his place:

Botany, Anatomy, Law, Geography, Geology, Literature, Travel Adventure, Economics, History, Chemistry, Agriculture, Accounting, Medicine, Pharmacy, etc. All in great quantity and in an astonishing variety.

But of all these works the most numerous by far are those of Mathematics, Physics, Chemistry and Biology.

My second observation concerned his practical equipment or what can be called laboratory equipment. These are slides in (Zoology, Botany, Water and Forests).

- *Three microscopes,*
- *A slide projector,*
- *A small 8mm film projector (animated)*
- *Two calculators*

I was also able to observe a collection of herbaria with more than 4,000 species, forming the basis of the national herbarium at the Faculty of Pharmacy and Botany of the UGANC.

Two jars containing snakes, several cards on Genetics, Agronomy, Geography, Periodic Classification of the Elements (MENDELEV table), Biology, Photo Synthesis,

Sex-linked Heredity, Cell Division, the Plant Kingdom etc.

My third observation concerned everything written by KARAMOKO. In this context and more than elsewhere, I realized that if he had had a little more time, he would have published a fairly large number of works, including:

a- Forty problems in the Baccalaureate

b- Inventory of natural resources

c- Bush fires

d- Rhizogenesis in the cutting

e- Tables of normal numbers

There was also :

- *A register containing KARAMOKO's reflections on man and the biosphere*
- *A register entitled Science containing reflections on the Biological Sciences*
- *Reflections on the Botany curriculum for Year 11*
- *Three records concerning the use of the herbarium:*

a- *Herbarium codes*

b- *Herbarium codes*

c- *Pharmacopoeia.*

- *A register entitled "Biological Evolution".*

- *Biochemistry*
- *Biological comments.*
- *Nimba biology composed as follows: a first part of the work devoted to 30 biology problems*
- *An excerpt from a reading entitled: Philosophy course.*
- *An analysis entitled: Entry to the Grandes Ecoles*
- *Collections such as: Science and Technology*

Science and technology

National Symposium on Folk Medicine Plants.

Science and literature.

- Agriculture :

Rice: Volumes 1 and 2

First mission in the brigades.

KARAMOKO had written a lot in other fields especially Mathematics for example.

- *Annales du Baccalauréat.*
- *Mathematics : (course)*
- *Mathesis.*
- *Applied mathematics*
- *100 practical problems (statements)*
- *100 practical problems (solutions)*
- *Mathematical logic and various problems*
- *Forestry mathematics*
- *Social mathematics*
- *Mathematics for agronomists*
- *Baccalaureate candidate's checklist*

- *Parcel differential equations*

He also made reading notes, four in number and all of which are literary reviews.

- *He also wrote about his travels, such as the trip to the DDR and the trip to China.*

- *At the end of the novels as "post 5" (Published by Harmattan Guinea ?)*

- *A translation from English: "In the country of Glélé (King of Dahomey)".*

As we speak, all of KARAMOKO's writings have been identified by a research mission and the Guinean Copyright Office.

KARAMOKO's contribution to Science

KARAMOKO was never satisfied with binding and writing abstract theories. More than the tranquility of an office, he especially loved the bush, manipulation, observation in the forests. He will walk all over Guinea.

At the complementary course of Macenta where he taught, all the students noticed very early his pragmatic rather than theoretical method. Almost all his classes were held outside.

In order to facilitate to the students the acquisition of the Latin names of the plants, KARAMOKO gave to each one of them a Latin name corresponding to that of a plant by saying to himself that a man cannot forget the name, the student will thus not have to forget the name of the plant.

*In 1966, together with his colleague **Fara CAMARA**, he succeeded in determining several plant species in Sérédou.*

*At Mount Nimba, he worked with one of his former students, **Jeremy KOMA**, who had also determined plants and owned a greenhouse in Kerouane.*

They will establish that these plants are xerophiles and can be used in drought control.

*It was also at Mount Nimba that, in the company of **Dr. Paul Condé, currently** Dean of the Faculty of Biology at the University of Conakry, he found the two main types of rare toads.*

a- ***The viviparous toad or Nectophrids***

b- ***Cane toad or Buffo occidentalis***

In Lola, they met his collaborator Lah Gonotey, a specialist in forestry sciences. He worked with him for some time.

When KARAMOKO returned to the bush, neither hunger nor thirst forced him to return home. In the bush, he ate wild edible fruits, which according to him, had therapeutic *values (preventive or curative). For lunch, in some places, he ate and drank decoctions and infusions of plants. In 1967 **with Mr. Stanislas LISSOVSKI** (Polish professor) he made study trips to observe the flora of the savannah. They were very often accompanied by the students of the IPK. These trips had such a strong impact on the student **Makan Kandé** that he specialized in plant systematics. He is currently working in Kankan.*

Still in Kankan, together with Professor CIUCA ILILILU (a Romanian) they made an entomological collection.

At the end of these outings, KARAMOKO was pleased to note that the flora of Kankan, Siguiri, Kouroussa and Kérouané, known as the flora of the savannah, is finally being studied.

This study will be added to those of Beyla, N'Zérékoré, Yomou, Lola, Macenta and Guéckédou.

When he was inspector of the Academy at N'Zérékoré, he was much more concerned with nature. It was also in N'Zérékoré that he began to think very seriously about procedures that could allow the experimental determination of plant names.

It was also in N'Zérékoré that he had to reflect on and find certain data concerning the chemical composition of plant secretions.

Thus, in 1969, Dr. Paul and Bah Ibrahima received from KARAMOKO two plant specimens whose names they did not retain. With these plants good results of chemical analysis, shops were obtained.

1- One of these plants soaked in an acid solution gave a red coloration and in a basic solution, blue.

2- The second one gave the opposite results.

The importance of this experiment is that it establishes that the barks of these plants could validly replace the difficult to convert indicator papers.

By indicator paper one expects in chemistry these papers which indicate the PH of a solution.

It is still in N'Zérékoré that he deepens his knowledge of plant drugs.

*During his stay in Conakry, in 1973-1975, as General Administrator of the IPGAN, he was undoubtedly occupied with the administration, but he devoted himself especially and completed his herbarium by the collection of plants of the zones of Conakry, Dubréka, Coyah, Forécariah and Boffa. Thus, **he transformed room 101 of the agronomy block into a herbarium. As a member of parliament, he wrote with Dr. M. M. the report on the state of the country.***

Kabinet Kanté a text on the importance of research for a developing country. This text, submitted to the National Assembly will be rejected because it is not understood in its essence.

Thereafter, KARAMOKO who did not like the Administration and hated politics above all, asked to be transferred to Faranah.

From Faranah, he will continue his work on the flora of Kissidougou, Faranah, Dabola and Dinguiraye.

*He often worked with **Mr. Maxime Lamotte and Mr. Schnell,** two great botanists known worldwide. In all the great international conferences of botany, these two professors will not cease to quote and to thank KARAMOKO for his contribution to science.*

*During his stay at IPGAN, he worked very often with **Mrs. Basilevskaia**. In Labé, he worked with a Soviet professor named **Kouzmine** on ornamental and vegetable plants.*

***Throughout his teaching life, his subjects proposed for the various examinations were always retained by the national commission**.*

KARAMOKO, thanks to his constant efforts and to the international conferences and colloquiums for which he was mandated by the Guinean government, managed to make himself known and admired by the great men of science. Thus he found himself representative of MAB in Guinea, member of the UMA (African Mathematical Union) and even member of the World Federation of Scientific Workers.

He participated in the protection of several classified forests in Guinea (Ziama in Macenta) and created a Botanical Garden in Faranah. He was one of the pioneers in the fight against bush fires in Guinea. In recognition of his numerous scientific and academic works, the Guinean government organizes every year a symposium on his life and works. A 'Karamoko KOUROUMA' prize for scientific research, of which I am the first laureate because of my work on alternatives to homologous transfusion (autotransfusion), is awarded each year by a jury to a researcher in Guinea.

This original work on termites and termite mounds, fruit of one of his numerous researches as a seasoned naturalist and which does not suffer from any conflict of interest, is the first of a series of numerous pending publications which I am making it my duty to improve, readapt and write for publication in scientific journals and books in order to pay a posthumous tribute to the man of whom I am the ^{third} son.

Acknowledgements

- *Mrs Hadja Fatoumata Keita Widow of Karamoko KOUROUMA*

- *Dr FAMORY KOUROUMA (his cousin), SOCIOLOGIST AND BIOGRAPHER at the University of Conakry*

- *Dr. Kabinet Kanté (mathematician)*

- *Pr Aliou Baniré Diallo (Mathematician, Coordinator of KK Scientific Circle)*

- *Messrs Maxime Lamotte and Schnell (naturalists)*

- *Mr. Stanislas LISSOVSKI (Polish professor)*

- *Mrs. Basilevskaia (Naturalist, Romanian Botanist)*

- *Makan Kandé (specialist in plant systematics)*

- *Dr. Paul Condé, currently Dean of the Faculty of Biology at the University of Conakry*

For their eloquent testimonies on the life and works of K. KOUROUMA and their participation, from near or far, in the various researches of the late Karamoko KOUROUMA

References

- *Polyphenism* in termites. Source Wikipedia -Termite
- Alba Zaremski, Daniel Fouquet and Dominique Louppe, *Les Termites dans le monde,* Ed. Quae, Versailles, 2009, 93 p. ISBN 978-2-7592-0343

- (Takuya Abe, David Edward Bignell and Masahiko Higashi (eds), *Termites: evolution, sociality, symbioses, ecology,* Kluwer Academic Publishers, Dordrecht; Boston; London, 2000, 466 p.) ISBN O-7923- 6361 (reprinted in 2002 in 2 volumes)

- (Krishna, Grimaldi, Krishna and Engel, 2013: *Treatise on the Isoptera of the world.* Bulletin of the American Museum of Natural History, [no.] 377, pp. 1-2704 (Full text)

- *Termites*. Lectures given by Alba Zaremski and Dominique Louppe on 10 May 2016 at the Nantes Natural History Museum. 18 pages report

Summary

I want morebooks!

Buy your books fast and straightforward online - at one of world's fastest growing online book stores! Environmentally sound due to Print-on-Demand technologies.

Buy your books online at
www.morebooks.shop

Kaufen Sie Ihre Bücher schnell und unkompliziert online – auf einer der am schnellsten wachsenden Buchhandelsplattformen weltweit! Dank Print-On-Demand umwelt- und ressourcenschonend produzi ert.

Bücher schneller online kaufen
www.morebooks.shop

KS OmniScriptum Publishing
Brivibas gatve 197
LV-1039 Riga, Latvia
Telefax: +371 686 204 55

info@omniscriptum.com
www.omniscriptum.com

MIX
Papier aus verantwortungsvollen Quellen
Paper from responsible sources
FSC® C105338

Printed by Books on Demand GmbH, Norderstedt / Germany